SCIENCE
A CLOSER LOOK

BUILDING SKILLS

Visual Literacy

Macmillan
McGraw-Hill

Instruction For Copying

Answers are printed in non-reproducible blue. Copy pages on a light setting in order to make multiple copies for classroom use.

Credits

Contents

LIFE SCIENCE

Chapter 1 KINGDOMS OF LIFE
Lesson 1 Cells .1
Lesson 2 Classifying Living Things 3
Lesson 3 The Plant Kingdom 5
Lesson 4 How Seed Plants
Reproduce 7

Chapter 2 THE ANIMAL KINGDOM
Lesson 1 Animals Without
Backbones 9
Lesson 2 Animals With Backbones 11
Lesson 3 Systems in Animals 13
Lesson 4 Animal Life Cycles 15

Chapter 3 EXPLORING ECOSYSTEMS
Lesson 1 Introduction
to Ecosystems 17
Lesson 2 Biomes 19
Lesson 3 Relationships
in Ecosystems 21

Chapter 4 SURVIVING IN ECOSYSTEMS
Lesson 1 Animal Adaptations 23
Lesson 2 Plants and Their
Surroundings 25
Lesson 3 Changes in Ecosystems 27

EARTH SCIENCE

Chapter 5 SHAPING EARTH
Lesson 1 Earth 29
Lesson 2 The Moving Crust 31
Lesson 3 Weathering and Erosion 33
Lesson 4 Changes Caused by the
Weather 35

Chapter 6 SAVING EARTH'S RESOURCES
Lesson 1 Minerals and Rocks 37
Lesson 2 Soil 39
Lesson 3 Resources from the Past 41
Lesson 4 Water 43
Lesson 5 Pollution
and Conservation 45

Chapter 7 WEATHER AND CLIMATE
Lesson 1 Air and Weather 47
Lesson 2 The Water Cycle 49
Lesson 3 Tracking the Weather 51
Lesson 4 Climate 53

**Chapter 8 THE SOLAR SYSTEM AND
BEYOND**
Lesson 1 Earth and Sun 55
Lesson 2 Earth and Moon 57
Lesson 3 The Solar System 59
Lesson 4 Stars and Constellations 61

PHYSICAL SCIENCE

Chapter 9 PROPERTIES OF MATTER
Lesson 1 Describing Matter 63
Lesson 2 Measurement 65
Lesson 3 Classifying Matter 67

Chapter 10 MATTER AND ITS CHANGES
Lesson 1 How Matter Can Change 69
Lesson 2 Mixtures 71
Lesson 3 Compounds 73

Chapter 11 FORCES
Lesson 1 Motion and Forces 75
Lesson 2 Changing Motion 77
Lesson 3 Work and Energy 79
Lesson 4 Simple Machines 81

Chapter 12 ENERGY
Lesson 1 Heat 83
Lesson 2 Sound 85
Lesson 3 Light 87
Lesson 4 Electricity 89
Lesson 5 Magnetism
and Electricity 91

How to Read Illustrations

In this workbook, you will learn how to read different kinds of illustrations. These include photos, diagrams, maps, tables, and charts. Charts and diagrams can help you understand what you are reading. They can also add information to what is written. In science, they might show how something works. They might show a sequence of events. Or they might show how things are different or alike.

Illustrations often have a title, caption, and labels. The title tells you what the illustration is about. The caption explains the illustration, or provides more information about it. If there are labels, they help to identify what is in the picture.

Humidity in a Rain Forest

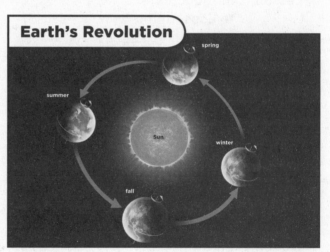

Earth's Revolution

The illustration on the left shows the rain forest. What can you learn about the rain forest by studying the photograph? The diagram on the right shows Earth's revolution. What do the arrows tell you about the direction of Earth's orbit and the seasons that result?

What are living things?

The table below lists five questions used to determine whether something is living or nonliving. If something performs all five life functions listed below, it is a living thing. If it does not, then it is a nonliving thing.

Answer these questions about the table.

Is It a Living Thing?			
Life Function	Lizard	Rock	Car
Does it grow?	✔	✘	✘
Does it use food to get energy?	✔	✘	✔
Does it get rid of wastes?	✔	✘	✔
Does it reproduce?	✔	✘	✘
Does it react to changes in its environment?	✔	✘	✘

1. Is a rock living or nonliving? How can you tell?

_____

_____

2. Compare and contrast cars and lizards. Which life functions do they carry out?

_____

_____

3. In what ways are lizards similar to humans?

_____

_____

4. Why do living things need in order to reproduce?

_____

_____

1. A rock is nonliving. It does not perform any of the five basic life functions.
2. Cars and lizards both use fuel and get rid of waste. Only lizards grow, reproduce, and respond to changes in their environments.
3. Both are living things that carry out the five basic life functions. Both are also made of cells.
4. Living things need to reproduce so that the group they belong to will survive.

Name _____ Date _____

How do plant and animal cells compare?

The table below compares and contrasts plant cells and animal cells. The cell parts of plants and animals are listed in the margin on the left. By reading each row from left to right, you can see if plant and animal cells both contain each cell part.

Answer these questions about the table.

1. Which type of cell has a cell wall?

_____

2. How do the cell drawings help to compare and contrast the cells?

_____

_____

3. What information does the table provide that is not evident in the drawings?

_____

_____

4. Why do plant cells usually have a green color and animal cells do not?

_____

_____

Cell Parts	Plant Cells	Animal Cells
cell wall	✔	✘
mitochondria	✔	✔
chloroplasts	✔	✘
nucleus	✔	✔
chromosomes	✔	✔
vacuole	large	small
cell membrane	✔	✔
cytoplasm	✔	✔

1. a plant cell
2. The drawings show the difference in shape and structure between plant and animal cells.

3. The table lists structures, such as vacuoles, that are not evident in the drawings.
4. Plant cells contain chlorophyll, which is green, and animal cells do not.

How are living things classified?

The chart below shows ways to classify living things.

Classifying Organisms

Kingdom	ancient bacteria	bacteria	protists	fungi	plants	animals
Number of Cells	one	one	one or many	one or many	many	many
Nucleus	no	no	yes	yes	yes	yes
Food	make their own or get food from other organisms	make their own or get food from other organisms	make their own or get food from other organisms	get food from other organisms	make their own food	get food from other organisms
Move from Place to Place	yes	yes	yes	no	no	yes

Answer these questions about the chart.

1. How are plants different from animals?

_____

2. How are fungi different from plants?

_____

3. Which kingdoms have animals that can make their own food?

_____

1. Plants make their own food but cannot move from place to place. Animals can move from place to place and get their food from other organisms.

2. They get their food from other organisms and can have only one cell.

3. Ancient bacteria, bacteria, protists, and plants can make their own food. Fungi and animals can only get their food from other organisms.

How are organisms named?

Both of the animals shown belong to the same genus.

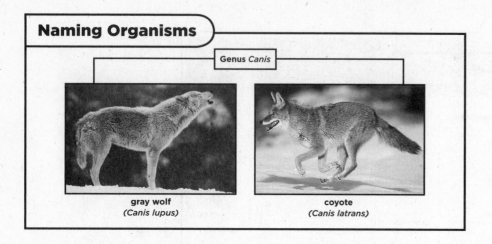

Naming Organisms

Genus *Canis*

gray wolf
(*Canis lupus*)

coyote
(*Canis latrans*)

Answer these questions about the diagram.

1. Which part of the coyote's scientific name refers to its genus? Which part of the name refers to its species?

   _____

2. What characteristics do these animals share?

   _____

3. A jackal has the scientific name *Canis aureus*. Would a photograph of a jackal fit in with the photographs above? Why or why not?

   _____

4. Why do you think scientists name organisms?

   _____

1. Canis refers to genus; latrans refers to species.
2. four legs, tails, body shape, and fur
3. Yes, because it belongs to the Canis genus.
4. Scientists name organisms to classify them.

© Macmillan/McGraw-Hill

Why are leaves important?

Plants get energy from the food they make in their leaves through photosynthesis. This diagram shows the process of photosynthesis. The arrows show the movement of sunlight, oxygen, carbon dioxide, water, and nutrients.

Answer these questions about the diagram.

1. What do the leaves take in during photosynthesis?

   _____

2. What do plants release during photosynthesis?

   _____

3. Where do water and nutrients go after they are taken in by the roots?

   _____

4. Put the following steps in order: (a) the plant gives off oxygen; (b) the plant takes in sunlight and carbon dioxide; and (c) the plant makes food.

   _____

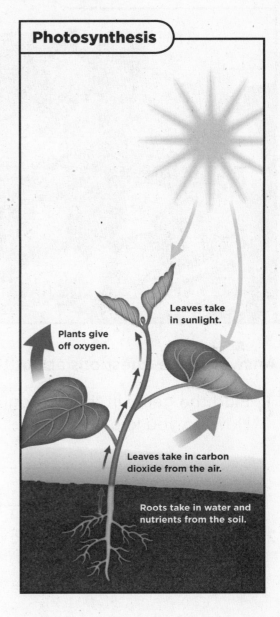

Photosynthesis

Leaves take in sunlight.

Plants give off oxygen.

Leaves take in carbon dioxide from the air.

Roots take in water and nutrients from the soil.

1. sunlight and carbon dioxide
2. oxygen
3. They are brought up the plant stem to the leaves.
4. b, c, a

How do we use plants?

People eat a variety of plant parts, including roots, leaves, stems, fruits, flowers, bulbs, and tubers.

Plants That People Eat

Answer these questions about the photograph.

1. Find the garlic in the photograph. Is it a bulb or a fruit? How do you know?

   _____

   _____

2. Suggest another title for the photograph.

   _____

   _____

1. Garlic is a bulb because it grows in the ground and does not have seeds.

2. Possible answers: Edible Plants and Living Things We Eat

How are plants alike and different from their parents?

Some traits, such as color, size, and shape, are inherited. This tree diagram shows how a farmer chose two specific pumpkins to create a larger pumpkin with specific traits.

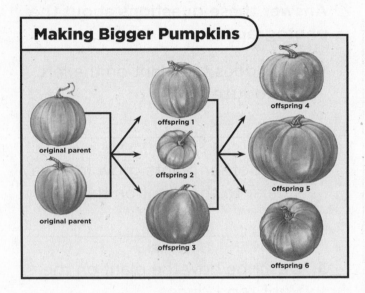

Making Bigger Pumpkins

original parent
original parent
offspring 1
offspring 2
offspring 3
offspring 4
offspring 5
offspring 6

Answer these questions about the diagram.

1. What trait was used to select the offspring to pollinate? Why?

_____

2. How many offspring did Offspring 1 and Offspring 3 produce?

_____

3. Which offspring was not pollinated? How do you know?

_____

1. Size—the largest pumpkins were selected, because that's the trait the farmer wanted.

2. three

3. Offspring 2; there is no bracket or arrow leading from Offspring 2 to another offspring.

Name _____ Date _____

What are other ways plants can reproduce?

Different types of plants reproduce in different ways. This photograph shows two of them. Notice the roots growing in the water without soil and the way the shoots grow from the bulb.

Reproduction Without Seeds

daffodil

chain plant

Answer these questions about the photograph.

1. How does the plant on the left reproduce?

_____

2. What other plants grow in a way similar to the plant on the left?

_____

3. Describe how the plant on the right reproduces.

_____

_____

4. What are other ways that plants can reproduce?

_____

_____

1. It grows from a bulb.
2. Tulips and onions also grow from bulbs.
3. The plant grows from cuttings. The cuttings can be placed in water to grow new plants.
4. Plants can reproduce using runners, flowers, cones, spores, seeds, and tubers.

What are invertebrates?

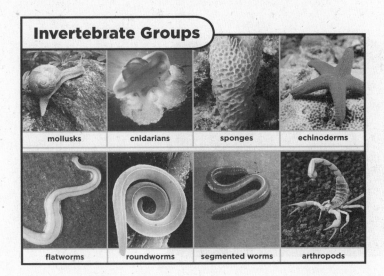

Invertebrate Groups

mollusks | cnidarians | sponges | echinoderms

flatworms | roundworms | segmented worms | arthropods

Answer these questions about the photographs.

1. In what way are all of these animals alike?

_____

2. The outer covering of the echinoderm appears to be spiny and hard. How is it different from the appearance of the cnidarian's outer covering?

_____

_____

3. How do photographs like these help scientists place animals into groups?

_____

_____

1. All of them are invertebrates.

2. It's the opposite. The cnidarian's outer covering appears to be smooth and soft.

3. These types of photographs help scientists compare and contrast different features of animals. Those with similar features are grouped together.

Name _____ Date _____

What are some invertebrates?

Before and After

Answer these questions about the photographs.

1. What are some physical features of this invertebrate?

_____

2. Which picture shows an octopus that is not being threatened? How can you tell?

_____

3. How is the appearance of the threatened octopus different from the non-threatened one?

_____

_____

1. It has tentacles, a large head, and a symmetrical body.

2. The picture on the left. The octopus appears to be relaxed.

3. The skin looks different, and its tentacles are curled up.

What are vertebrates?

Charts display information in an organized way. This chart shows some of the different vertebrate groups.

Answer these questions about the chart.

1. What physical feature do all of these vertebrates share?

_____

_____

_____

2. How are horses and birds similar?

_____

_____

_____

3. How are the skeletons of the shark and the horse different?

_____

_____

_____

Classes of Vertebrates

Cold-blooded

cartilaginous fish

reptiles | jawless fish

amphibians | bony fish

warm-blooded

birds | mammals

1. All of these vertebrates have a backbone.

2. They are both warm-blooded animals.

3. Sharks have skeletons made of cartilage. Horses have bony skeletons.

What are some other vertebrate groups?

Amphibians and reptiles look somewhat alike.
However, they have important differences.
Look at their bodies
and their environments.
How are they different?

Amphibians and Reptiles

lizard

frog

Answer these questions about the photographs.

1. To which animal group do frogs belong? What clue in
the photograph helps you answer this question?

_____

_____

_____

2. Describe the lizard's environment. How does this
environment help the lizard meet its needs?

_____

_____

1. Frogs are amphibians. The clue is that
the frog in the photograph is resting in
a pond. Amphibians have moist skin and
must live in or near water.

2. The lizard is in a warm, dry place. It
needs sunshine and warmth to control
its body temperature.

© Macmillan/McGraw-Hill

How do air and blood travel in the body?

Use these diagrams to compare the body systems of two different animals.

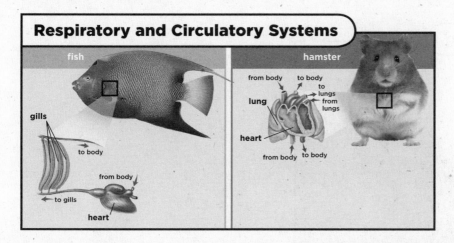

Respiratory and Circulatory Systems

fish

hamster

gills

to body

from body

to gills

heart

from body   to body
                        to lungs
                        from lungs
lung

heart

from body   to body

Answer these questions about the diagrams.

1. Which part of the circulatory system is illustrated in the diagrams?

_____

2. Which of the hamster's organs performs a function similar to that of the fish's gills?

_____

3. How many chambers are in the fish's heart? How many are in the hamster's heart? What is the heart's role in the circulatory system?

_____

1. The heart is the part of the circulatory system illustrated in the diagrams.

2. The lungs of the hamster perform a function like that of the fish's gills.

3. The fish's heart has two chambers. The hamster's heart has four chambers. The heart pumps blood to all parts of the body.

Name _____   Date _____

How is food broken down?

This diagram shows what a turtle would look like if you could see through its underside.

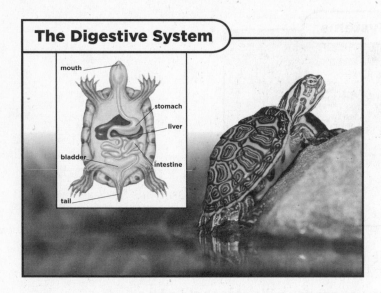

The Digestive System

- mouth
- stomach
- liver
- bladder
- intestine
- tail

Answer these questions about the diagram.

1. What system is illustrated in the diagram?

_____

2. What is the main function of this system?

_____

3. How does food change as it moves through the turtle's system?

_____

_____

1. The system in the diagram is the digestive system.

2. The main function of the digestive system is to break down food to get energy.

2. Food is gradually broken into smaller pieces and mixed with chemicals until the particles are small enough to be absorbed into the circulatory system.

What is metamorphosis?

Study the process of metamorphosis by looking at the photographs from left to right.

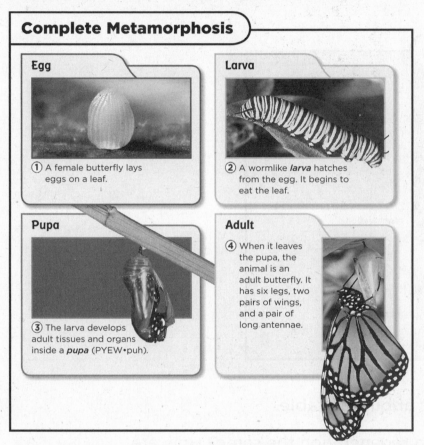

Complete Metamorphosis

Egg
① A female butterfly lays eggs on a leaf.

Larva
② A wormlike *larva* hatches from the egg. It begins to eat the leaf.

Pupa
③ The larva develops adult tissues and organs inside a *pupa* (PYEW•puh).

Adult
④ When it leaves the pupa, the animal is an adult butterfly. It has six legs, two pairs of wings, and a pair of long antennae.

1. A complete metamorphosis is shown. I can tell because there are four separate stages.
2. The four stages of metamorphosis, in order, are: egg, larva, pupa, adult.

Answer these questions about the photographs.

1. What type of metamorphosis is illustrated? How can you tell?

_____

_____

2. Name the four stages of metamorphosis in the order that they happen.

_____

_____

Name _____ Date _____

How do animals reproduce?

This table compares different ways of reproducing. The column on the right shows the number of parents needed.

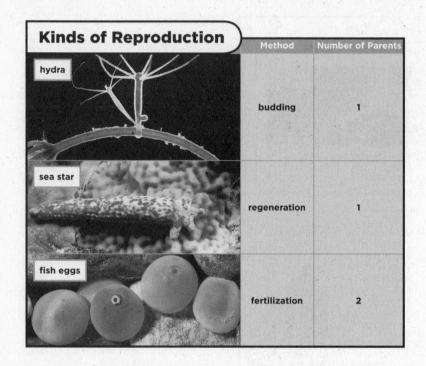

Kinds of Reproduction	Method	Number of Parents
hydra	budding	1
sea star	regeneration	1
fish eggs	fertilization	2

1. Both only require one parent. In budding, a bud forms on the adult and eventually develops into a new organism. In regeneration, a whole animal develops from just one part of the original animal.

2. Budding and regeneration involve one parent, while fertilization involves two parents.

Answer these questions about the table.

1. How are budding and regeneration the same? How are they different?

_____

_____

_____

2. How are budding and regeneration different from fertilization?

_____

_____

What is an ecosystem?

This diagram shows how the organisms in an ecosystem interact with each other and with the nonliving things in their environment.

A Pond Ecosystem

Answer these questions about the diagram.

1. What are two biotic factors in this ecosystem that frogs depend on?

   _____

   _____

2. Name two abiotic factors in this ecosystem that frogs need in order to survive.

   _____

   _____

3. How do frogs help the ecosystem?

   _____

   _____

1. Frogs need flies and plants for food and shelter.
2. Frogs need water and a warm climate.
3. Frogs help the ecosystem by eating flies. Frogs and their eggs can be eaten by other animals.

Name _____ Date _____

What are populations and communities?

Photographs can provide details about an environment. Compare the photographs to learn about the kinds of communities these ecosystems would support.

Populations and Communities

Answer these questions about the photographs.

1. What kind of ecosystem is shown in the photograph on the left?

   _____

2. What can you tell about this ecosystem from the photograph?

   _____

   _____

3. What ecosystem can you see in the photograph on the right?

   _____

1. A tropical rain forest is shown.

2. It has many plants and trees. There are also birds, so there must be insects and worms for them to eat. It also has water from a lake or river. The water is dark, so it must rain a lot.

3. a grassland ecosystem

© Macmillan/McGraw-Hill

What is a biome?

This map shows which biomes are found in North America and South America. The map key tells where the different types of biomes are found.

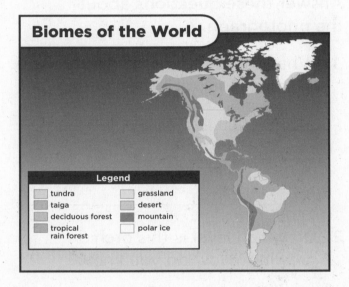

Biomes of the World

Legend
- tundra
- taiga
- deciduous forest
- tropical rain forest
- grassland
- desert
- mountain
- polar ice

Answer the questions about the map.

1. What are the names of the six biomes?

   _____

   _____

2. Where on this map would you expect to find tropical rain forests? Why?

   _____

   _____

3. Which biomes are found in the northern part of North America? Why?

   _____

   _____

Name _____ Date _____

What are grasslands and forests?

The photograph shows an orchid in its natural environment.
Where is it growing?

Tree Orchids

Answer these questions about the photograph.

1. What are the three main sections of the rain forest?

   _____

   _____

   _____

2. On which level is this orchid growing? How do you know?

   _____

   _____

   _____

   _____

   _____

   _____

1. the canopy, the understory, and the forest floor

2. The orchid is in the understory. The photograph shows that its roots are forming on the trunk of a tree. If it were growing in the canopy, there would be more sunlight and more leaves. If it were growing on the forest floor, it would be darker.

What is a food chain?

Read the diagram from top to bottom to understand how living things in a community depend on one another.

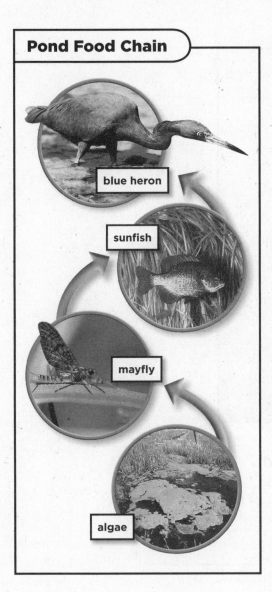

Pond Food Chain

blue heron

sunfish

mayfly

algae

Answer these questions about the diagram.

1. Are there producers in this diagram? How do they get energy?

_____

_____

_____

2. Are there consumers in this diagram? Are they herbivores or carnivores?

_____

_____

_____

1. Yes; the algae are producers. They get energy from the Sun.
2. Yes. The mayfly is an herbivore. The heron and the sunfish are carnivores.

Name _____ Date _____

What is a food web?

This diagram shows an ocean food web. Follow the arrows to see the relationships between the animals.

Ocean Food Web

Answer these questions about the diagram.

1. Which animals serve as prey for whales? How do you know?

   _____

   _____

2. Can an organism be both a predator and prey? Give an example from the diagram.

   _____

   _____

What are adaptations?

Look at the animals' bodies to see how they are adapted to their environments.

Animal Adaptations

arctic fox

fennec fox

Answer these questions about the photographs.

1. How is the arctic fox adapted to its environment?

_____

2. How is the fennec fox adapted to the desert?

_____

3. What might happen if the arctic fox were relocated to the desert?

_____

_____

1. The arctic fox has thick fur that is light in color to blend in with its environment.

2. It has short hair to keep cool, and its fur color helps it camouflage itself in the sand.

3. Its white fur would make it stand out, making it easy prey. Its thick fur would make it difficult to stay cool.

Name _____ Date _____

What are some other adaptations of animals?

This photograph shows a snow leopard in its environment. Compare its fur and markings to the colors in the environment.

Camouflage

Answer these questions about the photograph.

1. What type of adaptation does the snow leopard have?

_____

2. Would this adaptation be helpful in a snowy environment? Why or why not?

_____

_____

1. camouflage
2. Yes. It would be well-adapted to a snowy environment because its colors and patterns would blend in with the snow, branches, and rocks.

How do plants respond to their environment?

This diagram shows plant tropisms. The tip of the plant shoot on the left is wrapped in foil. Look for the differences in the way the two plants' shoots are growing.

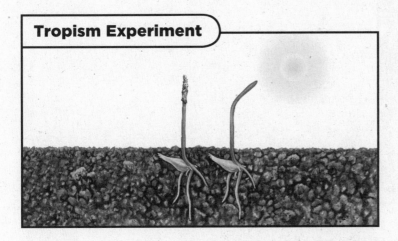

Tropism Experiment

Answer these questions about the diagram.

1. How are these plant shoots responding to the light?

_____

2. Why are the two plants responding differently?

_____

_____

3. What type of tropism is the second plant demonstrating?

_____

1. The first shoot is growing straight up. The second shoot is bending toward a light source.

2. The part of the plant that responds to

light is the tip of the shoot. The tip of the first plant shoot is covered with foil, so it does not respond.

3. light tropism

Name _____ Date _____

What are some plant adaptations?

This photograph was taken extremely close to the cactus so you can see its outer covering. Think about how the covering helps it to survive in its environment.

Desert Adaptations

Answer these questions about the photograph.

1. How is this prickly pear cactus different from most plants?

_____

2. Where would you expect to see this type of plant? Why?

_____

3. How does the outer covering of the cactus help it survive in its environment?

_____

1. It has a thick, waxy outer skin and needles.

2. I would expect to see it in a desert or other dry environment, because it has spongy tissue that stores water.

and waxy skin to prevent water from escaping.

3. The outer skin helps it hold in moisture, and the needles keep animals away.

What causes an ecosystem to change?

These photographs show the same volcano and the surrounding environment at two different points in time. Look for differences between the photographs.

Natural Change in Ecosystems

Mount Saint Helens in 1980

Mount Saint Helens in 1995

Answer these questions about the photographs.

1. Why is the land barren in the first photograph?

_____

2. How has the ecosystem changed in the second photograph?

_____

3. If you saw a photograph of the same area taken in 1979, what might it look like? Explain.

_____

_____

1. The volcano erupted, and the hot lava killed the plant life.
2. Plant life has grown on and around the volcano.

3. The ground may have been covered with plant life before the volcano erupted. If there were earlier eruptions, it may look the same as in the first photograph.

Name _____   Date _____

How do people change ecosystems?

The photographs below show how people can change underwater ecosystems.

Manufactured Ecosystems

an airplane being sunk

a subway car being sunk

Answer these questions about the photographs.

1. How are the plane and subway car being recycled?

_____

2. Why do barrier reefs need to be replaced?

_____

_____

1. They are being used to create artificial barrier reefs.

2. The reefs are important, because they are home to many organisms. Without the reefs, these organisms could not survive in this area.

What does Earth's land look like?

A relief map displays the shape of the land. This relief map shows some of Earth's landforms that make up the United States. The raised areas are hills and mountains. The flat areas are plains.

The Continental United States

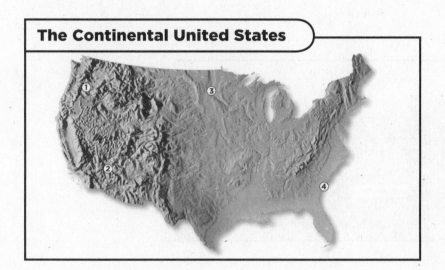

1. the western part, or the West
2. the central part or Midwest; a plain
3. The central part of the country is best for farming, because the land is flat and made up of mostly plains.
4. Plains and mountains are found in the Northeast.

Answer these questions about the map.

1. Which part of the country has many mountains?

_____

2. Which part of the country is mostly flat? What is the flattest landform called?

_____

3. What area of the country is best for farming? Explain your answer.

_____

_____

4. What types of landforms are found in the Northeast?

_____

Name _____ Date _____

What does it look like where water meets land?

Have you ever wondered what the ocean floor looks like? This diagram shows features of a large piece of the ocean floor in deep water.

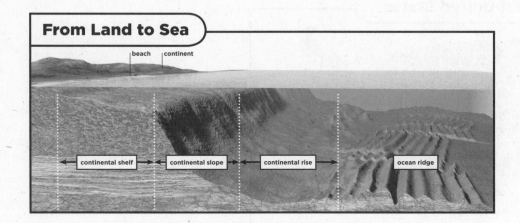

From Land to Sea

beach   continent

continental shelf → continental slope → continental rise   ocean ridge

1. The continent has a higher elevation.
2. The continental slope is steepest.
3. The continental rise connects the continent to the ocean floor.
4. The water eroded the surface of the ocean floor.

Answer these questions about the diagram.

1. Which area has a higher elevation—the continent or the ocean floor?

_____

2. Which part of the ocean floor is steepest?

_____

3. What area connects the continent to the ocean floor?

_____

4. What do you think shaped the ocean floor?

_____

How does Earth's crust move?

Earth's crust changes based on its surroundings. These
cross-sections illustrate slices of Earth's plates. They show
the plate motions that cause different mountain formations.

Mountains in the Making

Answer these questions about the diagram.

1. Which part of the diagram shows a fold mountain? How
do you know?

_____

_____

2. Which part of the diagram shows the formation of a
fault-block mountain? How do you know?

_____

_____

1. The part on the right shows a fold
mountain. Fold mountains form where
plates meet at the edges of continents,
causing the land to scrunch up between
them. It shows the folding of Earth's
surface as these plates meet.

2. The part on the left shows the
formation of a fault-block mountain.
It shows the fault that is made by the
movement of one plate sliding up and
another plate sliding down.

© Macmillan/McGraw-Hill

How do scientists study earthquakes?

Scientists study past events in order to predict future events. This timeline shows tools people have used to measure earthquakes.

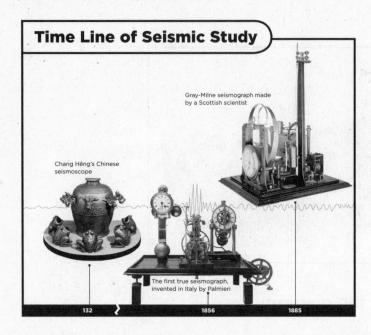

Time Line of Seismic Study

Gray-Milne seismograph made by a Scottish scientist

Chang Hĕng's Chinese seismoscope

The first true seismograph, invented in Italy by Palmieri

132   1856   1885

Answer these questions about the diagram.

1. How do the tools change as you read the timeline from left to right?

_____

2. How many years passed between Heng's seismoscope invention and Palmieri's seismograph invention? How does it compare to the amount of time between Palmieri's seismograph invention and Gray-Milne's seismograph invention?

_____

1. The tools become more complicated and more technologically advanced from left to right.

2. There were 1,724 years between Heng's and Palmieri's inventions, and

29 years passed between Palmieri's and Gray-Milne's inventions. More time passed between Heng's and Palmieri's inventions than between Palmieri's and Gray-Milne's inventions.

What is erosion?

The photograph below shows the Colorado River running through the Grand Canyon. It also illustrates the effects of weathering. Look carefully at the course and location of the river. Study the shading and texture of the cliffs.

Rivers Shape the Land

Answer these questions about the photograph.

1. What evidence of erosion do you see in the photograph?

_____

2. What do you think caused the erosion?

_____

3. What details suggest that it took thousands of years for the Grand Canyon to form?

_____

1. The sides of the canyon are steep and worn down.

2. The river caused the erosion by wearing down the sides of the rock over time, creating the canyon.

3. The many layers of rock and the depth of the canyon show that the Grand Canyon was formed over a long period of time.

Name _____ Date _____

How do glaciers shape the land?

The diagram below shows how glaciers move and deposit land. Because glaciers move downhill, you should begin reading the diagram in the upper left corner and follow it to the bottom right corner of the diagram.

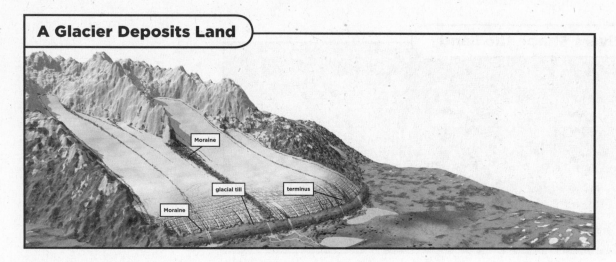

A Glacier Deposits Land

Moraine

glacial till

terminus

Moraine

Answer these questions about the diagram.

1. How does glacial debris end up at the bottom of a glacier?

_____

_____

2. How would the land at the top of the diagram change if the glacier moved across it? How would the land change at the bottom?

_____

_____

1. When some of the glacier melts, it begins to move, picking up rocks and anything else in its path. The glacier then carries much of the glacial debris downhill.

2. The land at the top of the diagram would wear away, leaving a wider, steeper valley in its place. The land at the bottom of the diagram would build up from deposited debris.

© Macmillan/McGraw-Hill

How do floods and fires change the land?

Before and After

Answer these questions about the photographs.

1. How does the picture on the left help you to understand how floods change the land?

_____

_____

2. What might have caused the flood?

_____

_____

3. What might a third photograph show if it were taken just after the floodwater dried up?

_____

_____

1. The picture on the left shows how the land looked before the flood.

2. Heavy rains or quickly melting snow might have caused a river to overflow, creating the flood.

3. A third photograph might show sediment on the pavement. Water might have washed away objects or moved them to other locations.

Name _____ Date _____

How do storms change the land?

Storms change the land in many ways. This aerial photograph shows what the land looked like from the sky after a hurricane passed through the southern United States.

Answer these questions about the photograph.

1. What does this aerial photo help you to see that a photo taken from the ground could not?

   _____

   _____

   _____

Hurricane Damage

2. What type of damage did the hurricane cause?

   _____

   _____

3. What characteristics of a hurricane could be responsible for this damage?

   _____

   _____

1. The photo taken from above allows me to see a larger area. This helps me to see the way the storm changed the whole region of land.

2. The hurricane damaged roofs, broke windows, and destroyed many of the houses in the area.

3. The strong winds and heavy rain.

© Macmillan/McGraw-Hill

What is a mineral?

Study the scale to compare and contrast the hardnesses of the minerals shown.

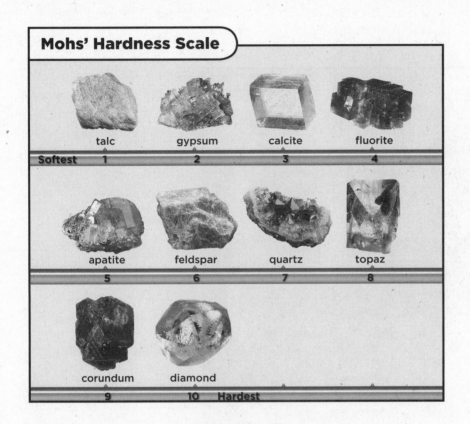

Mohs' Hardness Scale

talc	gypsum	calcite	fluorite
Softest 1	2	3	4
apatite	feldspar	quartz	topaz
5	6	7	8
corundum	diamond		
9	10 Hardest		

1. Hardness measures a mineral's ability to scratch or to be scratched by another mineral.
2. corundum and diamond
3. talc

Answer these questions about the table.

1. What does hardness measure?

_____

_____

2. Which minerals can scratch topaz?

_____

3. Which mineral cannot scratch any of the other minerals shown?

_____

Name _____ Date _____

What are metamorphic rocks?

The diagram below shows how rocks change from one type to another. Follow the arrows to help you understand the rock cycle.

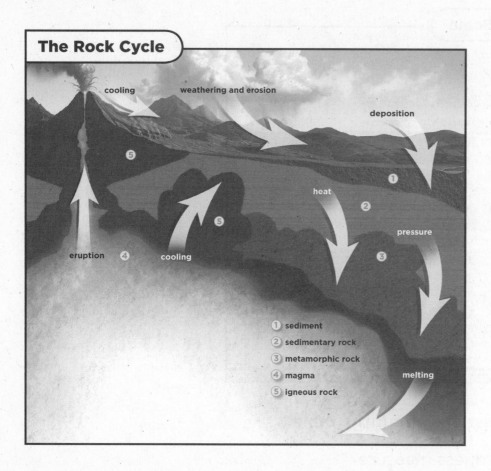

The Rock Cycle

cooling   weathering and erosion

deposition

⑤

heat   ①

⑤   ②

pressure

eruption   ④   cooling   ③

① sediment
② sedimentary rock
③ metamorphic rock
④ magma   melting
⑤ igneous rock

1. Igneous rocks are formed from melted sedimentary and metamorphic rocks or by cooled lava.

2. Rocks are transformed by heat and pressure.

Answer these questions about the diagram.

1. How do igneous rocks form?

_____

_____

2. How do rocks transform from one type into another?

_____

_____

© Macmillan/McGraw-Hill

What is soil made of?

The photograph below shows an example of how an animal can cause weathering.

Weathering Caused by Living Things

Answer these questions about the photograph.

1. How does soil form?

   _____

   _____

2. What is the animal in the photograph doing? How does this help create soil?

   _____

   _____

3. What are other ways that living things help to produce soil?

   _____

   _____

1. Soil is created as bedrock weathers and breaks into smaller and smaller pieces.
2. The animal is burrowing or digging a hole. This helps to break the soil into smaller pieces.
3. Plant roots can break apart rocks. When living things die, they decompose and form humus.

Name _____ Date _____

What are some properties of soil?

The diagram below includes a magnified view of the soil to help you observe the soil's properties.

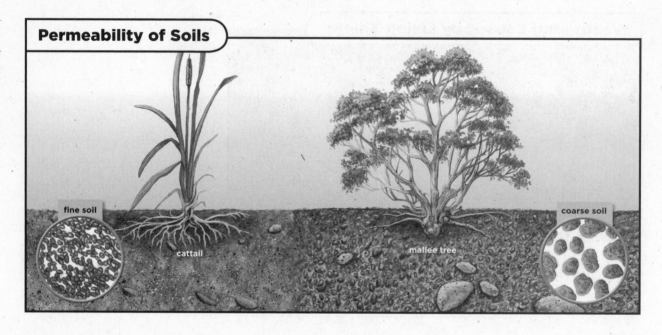

Permeability of Soils

fine soil

cattail

mallee tree

coarse soil

Answer these questions about the diagram.

1. What are pore spaces? What do they indicate about soil?

_____

_____

2. Do you think that this soil is very permeable? How can you tell?

_____

_____

1. Pore spaces are the spaces between particles of soil. The size and number of the pore spaces determine how porous the soil is.

2. Yes; the pore spaces are large and connected.

What are fossil fuels?

Much of the energy that we use comes from fossil fuels. Look at the steps in the diagram below to see how fossil fuels are formed.

How Coal Forms

① Dead plants sink to the bottom of a swamp.

② A thick layer of decaying plants builds up.

③ Decaying plants become part of a sedimentary rock layer.

④ The rock layer is pressed into soft coal.

⑤ Under intense heat and pressure, the soft coal turns to hard coal. It is now a fossil fuel.

Answer these questions about the diagram.

1. What makes up soft coal?

_____

2. What turns soft coal into hard coal?

_____

3. How long does it take to form fossil fuels?

_____

4. How are fossil fuels obtained?

_____

1. decaying plants and sedimentary rock   3. millions of years
2. intense heat and pressure   4. by mining and digging deep below the Earth's surface

What can we use instead of fossil fuels?

Fossil fuels are a nonrenewable source of energy. The Earth and the Sun provide us with renewable sources, as well. The chart below shows the percentage of renewable and nonrenewable sources of electricity that we used in 2005.

Answer these questions about the chart.

1. What percentage of electricity came from renewable energy in 2005?

   _____

   _____

2. Did more electricity come from renewable or nonrenewable resources?

   _____

   _____

3. Is the electricity generated from wind mills, solar power, and ocean tides renewable or nonrenewable energy?

   _____

   _____

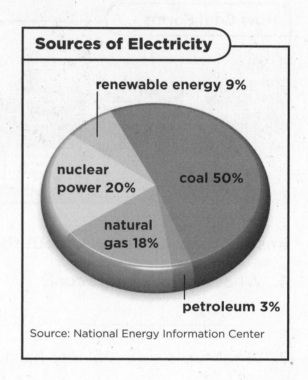

Sources of Electricity

renewable energy 9%

nuclear power 20%

coal 50%

natural gas 18%

petroleum 3%

Source: National Energy Information Center

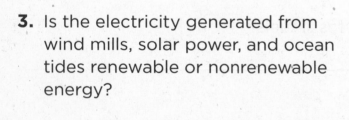

1. 9 percent
2. nonrenewable resources
3. renewable energy

Where is Earth's water found?

This map shows a major waterway. Follow the dotted lines from Lake Superior to the Gulf of St. Lawrence.

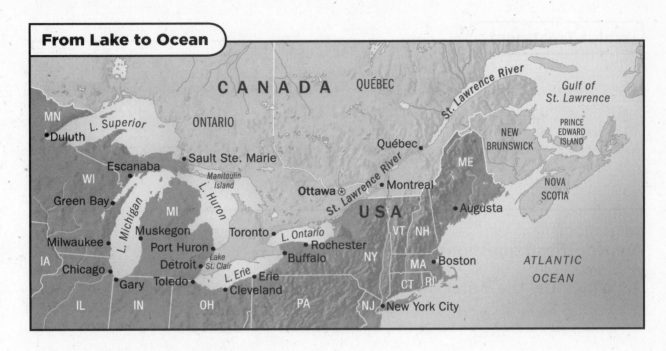

From Lake to Ocean

Answer these questions about the map.

1. What route could a ship take from Cleveland to the Atlantic Ocean?

   _____

   _____

2. Where might runoff water from Ottawa end up?

   _____

   _____

1. The ship could travel across Lake Erie and Lake Ontario. Then it could travel across the St. Lawrence River to the Gulf of St. Lawrence and into the Atlantic Ocean.

2. Runoff may flow into the St. Lawrence River, which flows into the Atlantic Ocean.

Name _____ Date _____

How is fresh water supplied?

Read the diagram below from left to right to understand how water goes from its source to the people who use it.

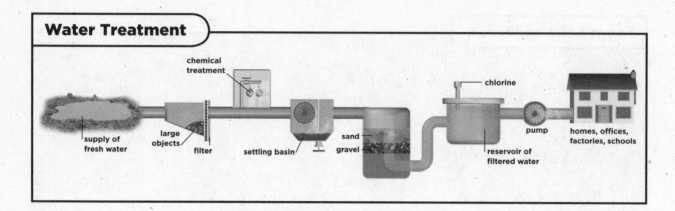

Water Treatment

chemical treatment

chlorine

supply of fresh water

large objects

filter

settling basin

sand gravel

reservoir of filtered water

pump

homes, offices, factories, schools

Answer these questions about the diagram.

1. Why must the water be treated?

_____

2. In what stage are tiny organisms killed?

_____

3. Where does the water travel when it leaves the chemical treatment basin? Explain how this next step helps to treat water.

_____

_____

1. It usually holds useless or even dangerous impurities.
2. chemical treatment
3. Water travels to a filtration reservoir, where sand and gravel are used to filter out impurities.

How can we protect soil and water?

Some farmers have always used contour plowing to grow better crops. Study the plow lines in the photograph below.

Contour Plowing

Answer these questions about the photograph.

1. How does contour plowing protect the soil?

   _____

   _____

2. How does contour plowing conserve water?

   _____

   _____

© Macmillan/McGraw-Hill

Name _____ Date _____

What are the 3 Rs?

The photographs below show examples of the 3 Rs. As you look at the photographs, think about other ways to conserve resources.

The 3 Rs in Action

WE RECYCLE

Answer these questions about the photographs.

1. Which photograph shows a material being reused? How is it reused?

   _____

   _____

2. How is reducing different from recycling?

   _____

   _____

1. The second photograph shows material being reused. An old tennis ball is being used as a home for animals.

2. To reduce means to use less of something, while to recycle is to make a new product from something that has already been used.

© Macmillan/McGraw-Hill

What is in the air?

The diagram below shows the layers of the atmosphere surrounding Earth.

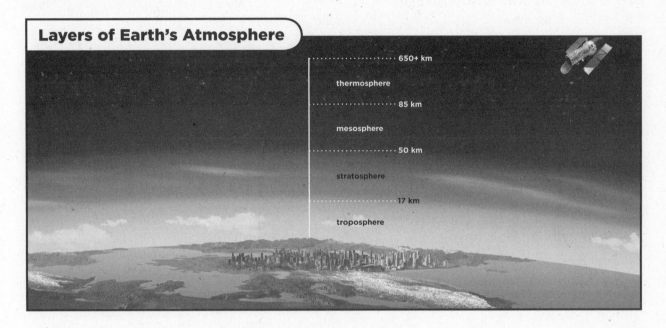

Layers of Earth's Atmosphere

- 650+ km
- thermosphere
- 85 km
- mesosphere
- 50 km
- stratosphere
- 17 km
- troposphere

Answer these questions about the diagram.

1. List, in order, the layers of the atmosphere that a rocket would pass through as it leaves Earth.

   _____

2. How thick is the stratosphere?

   _____

3. In which layer might you find a satellite?

   _____

1. troposphere, stratosphere, mesosphere, thermosphere
2. It's 33 km thick.
3. thermosphere

Name _____ Date _____

What are some properties of weather?

The photograph below shows the abundance of plants in a rain forest. Think about the type of weather conditions that are shown in this photograph.

Humidity in a Rain Forest

Answer these questions about the photograph.

1. What can you infer about the temperature in the picture? How can you tell?

_____

_____

2. What factors make the rain forest humid?

_____

_____

3. What could be added to this rain forest to increase the humidity?

_____

_____

1. It must be high. It's hazy and sunlight is shining through the leaves.
2. The high temperature and water vapor in the air make it humid.
3. a body of water, like a lake or river

© Macmillan/McGraw-Hill

Where does water go?

Read the captions and follow the arrows on the diagram to understand the phases of the water cycle.

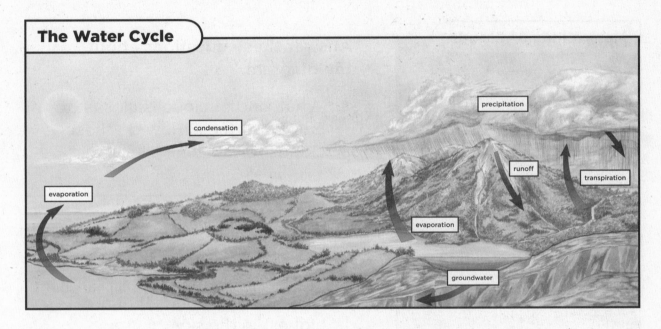

The Water Cycle

condensation

precipitation

runoff

transpiration

evaporation

evaporation

groundwater

Answer these questions about the diagram.

1. What happens when the Sun heats the surface of the ocean?

_____

2. How are clouds formed?

_____

_____

3. How does water return to the ocean?

_____

1. Water evaporates, and as it rises, it becomes water vapor.

2. As the water vapor rises, it cools, condenses, and forms clouds.

3. It returns to the ocean as precipitation. When it rains, it goes directly into the ocean, or it falls into rivers and then empties into the ocean.

What are some types of clouds?

Look at the water droplets on the web. Think about where the water came from and where it might go.

Many Kinds of Clouds

cirrus

cirrocumulus

altocumulus

cumulonimbus

Answer these questions about the diagram.

1. What kind of cloud is closest to Earth's surface? How do these clouds form?

   _____

   _____

   _____

2. Describe the clouds that are highest in the sky.

   _____

   _____

   _____

3. What kind of cloud looks like thick fog?

   _____

   _____

   _____

1. The cumulonimbus cloud is closest. They form when cumulus clouds get darker and thicker.

2. The cirrus clouds are highest in the sky.

They are thin, feathery, and contain tiny bits of ice.

3. The stratus cloud looks like thick fog.

What are air masses and fronts?

This diagram shows the fronts that form between separate air masses.

Different Fronts

warm

cold

warm front

warm

cold

cold front

cold   warm

stationary front

Answer these questions about the diagram.

1. Which kind of front brings thick clouds and stormy weather?

_____

_____

2. Which kind of front brings rainy weather?

_____

_____

3. What is the difference between a warm front and a stationary front?

_____

_____

1. A cold front brings stormy weather.

2. A warm front brings rainy weather.

3. When a warm air mass overtakes a cold air mass, a warm front forms. There's light but steady rain until it passes, and then the temperature gets warmer. When the air masses are not moving, a stationary front forms. This may bring rainy weather that lasts for days.

Name _____ Date _____

What does a weather map show?

On this map, one key tells the symbols for fronts, pressure, and types of precipitation. The other one uses colors to indicate the range of temperatures.

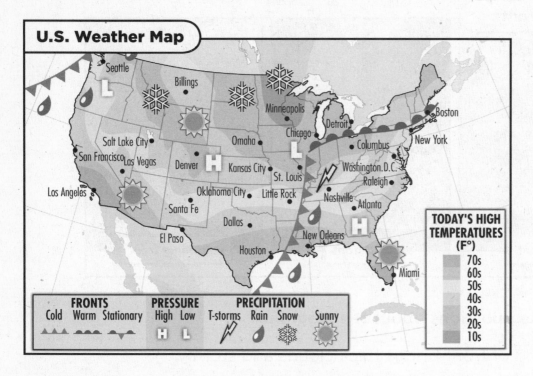

U.S. Weather Map

TODAY'S HIGH TEMPERATURES (F°)
70s
60s
50s
40s
30s
20s
10s

FRONTS — Cold Warm Stationary
PRESSURE — High Low H L
PRECIPITATION — T-storms Rain Snow
Sunny

Answer these questions about the map.

1. Near what midwestern city is the air pressure low?

   _____

2. What type of weather has formed along the cold fronts on this map?

   _____

3. What type of weather might a forecaster predict for New York for the next few days? Why?

   _____

   _____

What determines climate?

Follow the arrows on this map to see how water moves in the ocean. The movement of ocean currents has a major impact on climate.

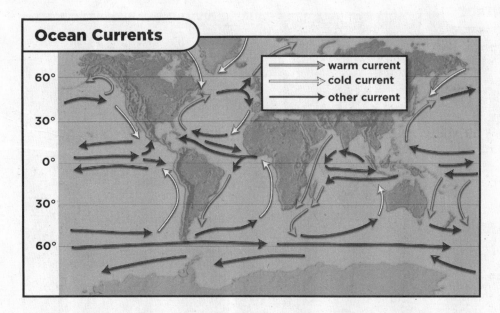

Ocean Currents

→ warm current
⇨ cold current
→ other current

60°
30°
0°
30°
60°

Answer these questions about the map.

1. How are the currents moving in the Pacific Ocean off the west coast of South America?

   _____

2. How does the current flow from the Gulf of Mexico off the southeastern coast of the United States?

   _____

3. Which coast has a warmer average temperature? Explain.

   _____

   _____

   _____

1. They are moving counterclockwise in a circular path.

2. northward, along the eastern coast of the United States

3. The East Coast has a warmer average temperature because there are more currents running north from the equator.

Name _____ Date _____

How do mountains affect climate?

The diagram shows how an air mass would move over the mountain in the photograph.

The Mountain Effect

An air mass loses moisture as it moves over a mountain.

Answer these questions about the photograph.

1. How does an air mass change as it travels over a mountain? Explain.

   _____

   _____

   _____

   1. It loses moisture. As it moves over the mountain, it rises and any water vapor condenses into clouds that turns into precipitation. That dries out the air mass.

2. What is the climate like on the right side of this mountain?

   _____

   2. It is drier.

© Macmillan/McGraw-Hill

What causes day and night?

The diagram below uses arrows and a model of Earth to show how Earth's rotation causes day and night.

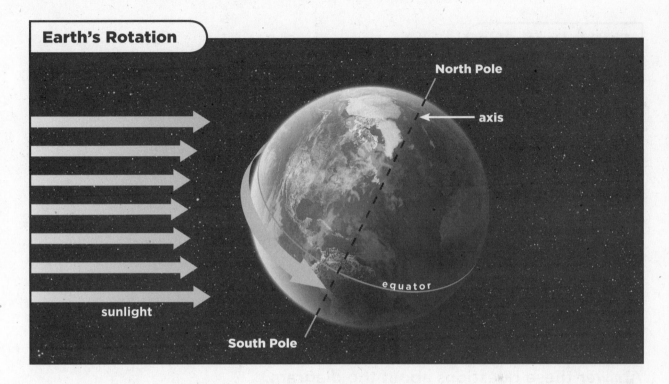

Earth's Rotation

North Pole

axis

equator

sunlight

South Pole

Answer these questions about the diagram.

1. Which side of Earth is experiencing daytime?

_____

2. How would this diagram be different if it were 12 hours later?

_____

3. How many hours does it take Earth to return to the same position?

_____

1. The western side of Earth is experiencing daytime.

2. The side that is experiencing daytime would be experiencing nighttime.

3. It takes Earth 24 hours.

Name _____ Date _____

What causes seasons?

Follow Earth's path as it revolves around the Sun. Notice how Earth's axis is tilted.

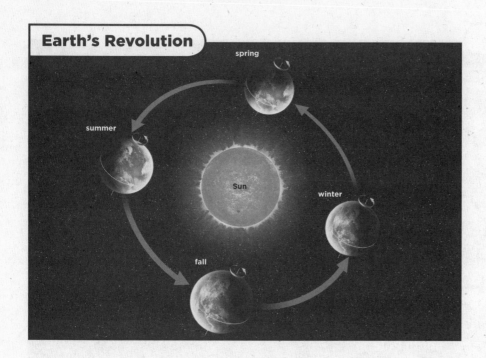

Earth's Revolution

spring

summer

Sun

winter

fall

1. The North Pole tilts away from the Sun. Sunlight strikes the Northern Hemisphere at low angles.
2. It is experiencing summer.
3. During fall, the Northern Hemisphere is tilted away from the Sun.

Answer these questions about the diagram.

1. What causes winter in the Northern Hemisphere?

_____

_____

2. If the Northern Hemisphere is experiencing winter, what is the Southern Hemisphere experiencing?

_____

3. Explain the position of the Northern Hemisphere during fall.

_____

_____

What are the phases of the Moon?

The larger moons in the diagram show how the Moon appears from Earth at different times of the month.

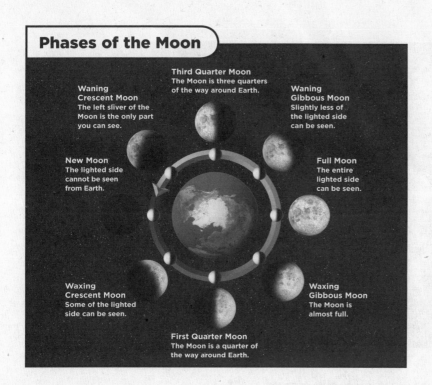

Phases of the Moon

Third Quarter Moon
The Moon is three quarters of the way around Earth.

Waning Crescent Moon
The left sliver of the Moon is the only part you can see.

Waning Gibbous Moon
Slightly less of the lighted side can be seen.

New Moon
The lighted side cannot be seen from Earth.

Full Moon
The entire lighted side can be seen.

Waxing Crescent Moon
Some of the lighted side can be seen.

Waxing Gibbous Moon
The Moon is almost full.

First Quarter Moon
The Moon is a quarter of the way around Earth.

1. A waxing Gibbous moon appears after the full Moon.

2. The left sliver of the Moon is the only part that is visible.

Answer these questions about the diagram.

1. What phase appears after the full Moon?

_____

_____

2. What part of the Moon is visible from Earth during a waning crescent Moon?

_____

_____

_____

Name _____ Date _____

What is an eclipse?

Compare the placement of the Sun, the Moon, and Earth to understand the difference between these eclipses.

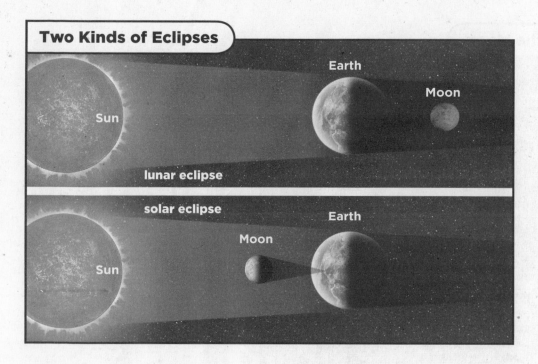

Two Kinds of Eclipses

Earth

Moon

Sun

lunar eclipse

solar eclipse

Earth

Moon

Sun

Answer these questions about the diagram.

1. What is an eclipse?

_____

_____

2. What happens during a lunar eclipse?

_____

_____

3. What causes a solar eclipse?

_____

_____

1. An eclipse is a shadow cast by Earth or the Moon.

2. Earth moves between the Sun and Moon and casts a shadow on the Moon.

3. A solar eclipse occurs when the Moon casts a shadow on Earth. This happens when the Moon is directly between the Sun and Earth.

© Macmillan/McGraw-Hill

What is the solar system?

This diagram shows the inner planets of our solar system.
The lines show the path of each planet's orbit around
the Sun.

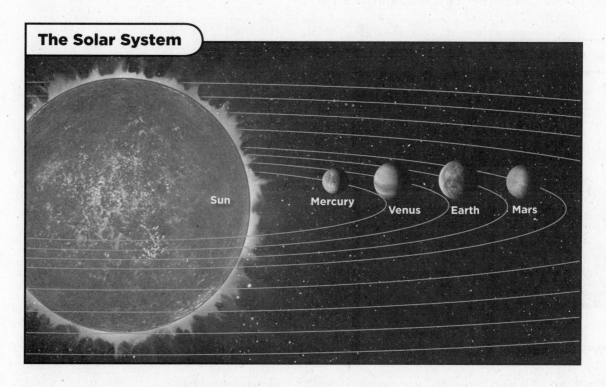

The Solar System

Sun   Mercury   Venus   Earth   Mars

Answer these questions about the diagram.

1. Describe the shape of the orbits.

_____

_____

2. What happens to the orbits as you get farther from
the Sun?

_____

_____

1. They are elliptical.   2. The orbits get larger.

How do we learn about the solar system?

The photographs below show how the study of the solar system has changed since Galileo's time.

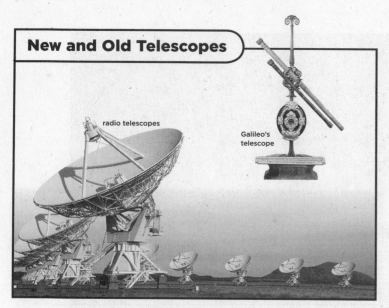

New and Old Telescopes

radio telescopes

Galileo's telescope

Answer these questions about the photographs.

1. How are these two telescopes similar? How are they different?

_____

_____

2. Where do radio telescopes appear to be located? Why do you think they are located here?

_____

_____

1. Both are used to study objects far away. The radio telescope lets scientists more clearly see things that are farther away.

2. They are in large, open areas, far from other objects. They can get a clearer view, and they take up a lot of room.

What are stars?

The scale at the bottom of this diagram shows the distance of some stars from Earth in light-years.

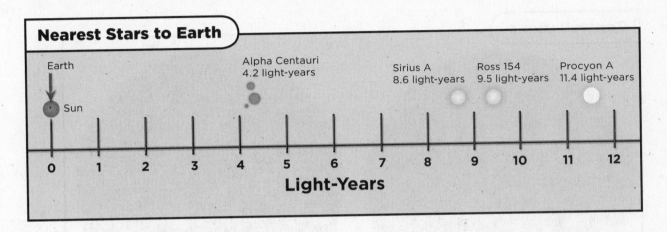

Answer these questions about the diagram.

1. How many light-years does it take for light from Sirius A to reach Earth?

_____

2. What is the distance between Procyon A and Earth?

_____

3. Which star is closest to Ross 154?

_____

_____

1. It takes 8.6 light-years for light from Sirius A to reach Earth.
2. The distance between Procyon A and Earth is 11.4 light-years.
3. Sirius A is closest to Ross 154.

Name _____ Date _____

What are constellations?

These star maps allow you to compare the constellations that can be seen in each hemisphere.

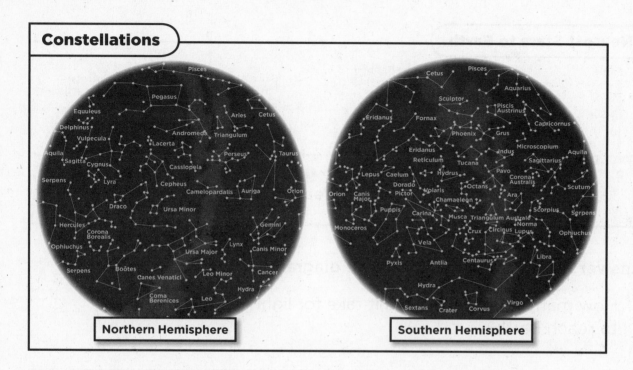

Constellations

Northern Hemisphere

Southern Hemisphere

Answer these questions about the diagram.

1. Why are the two star maps different?

_____

_____

2. Would the constellations look the same if viewed from Mars? Explain.

_____

_____

1. They look different because constellations depend on your position on Earth. The night sky looks different in the Northern Hemisphere than it does in the Southern Hemisphere.

2. No, because the viewing position in Mars in relation to the stars would be different than it is on Earth.

What is matter?

This photograph shows a balance scale. Look at the objects on the scale. Think about how the scale is used to measure mass.

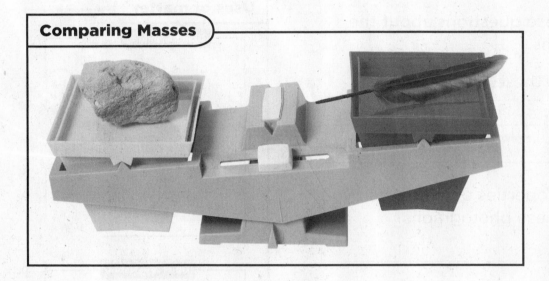

Comparing Masses

Answer these questions about the photograph.

1. Which of these objects has the greater mass?

_____

2. How would the scale look if the objects had the same mass?

_____

3. What unit of measurement would be used to measure the mass of the rock?

_____

1. The rock has a greater mass than the feather.

3. Grams would be used.

2. The scale would show the objects at the same level if they had the same mass.

Name _____ Date _____

What happens to the matter we use?

Look at the photographs. These objects have been classified into two sets. Think about the properties that are shared by members of each group.

Answer these questions about the photographs.

1. How are these items classified?

_____

_____

2. What properties of matter can you see in these photographs?

_____

_____

3. How would you find the buoyancy of these objects?

_____

_____

Uses of matter

Objects Made by People

Objects in Nature

© Macmillan/McGraw-Hill

1. The items in the first set were made by people, and the items in the second set were made by nature.

2. I see color, size, and shape.

3. To find the buoyancy, I would place each object in water and see whether or not it floats. If it floats, then it is buoyant.

How do we measure matter?

Read the table from left to right. Note the common objects listed beside the metric measurements.

Metric Units	Amount	Estimated Length
1 centimeter (cm)	$\frac{1}{100}$ of a meter	the width of your thumbnail
1 decimeter (dm)	10 cm $\frac{1}{10}$ of a meter	the length of a crayon
1 meter (m)	10 dm 100 cm	the length of a baseball bat
1 kilometer (km)	1,000 m 100,000 cm	the distance you walk in 10 to 15 minutes

Answer these questions about the table.

1. How many meters are in a kilometer?

_____

2. Place these units in order from smallest to largest: kilometer, decimeter, centimeter, and meter.

_____

3. What unit of measurement would most likely be used to measure the length of a racetrack?

_____

_____

_____

1. There are 1,000 meters in a kilometer.
2. From smallest to largest, the units are: centimeter, decimeter, meter, kilometer.
3. Kilometers would most likely be used to measure the length of a racetrack.

Name _____ Date _____

What is density?

This diagram shows the difference between the air particles in the balloon and those in the air.

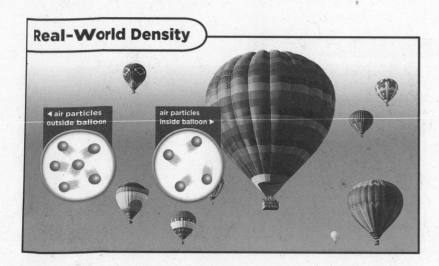

Real-World Density

◄ air particles outside balloon

air particles inside balloon ►

Answer these questions about the diagram.

1. Look at the balloons. How does the density of air particles inside the balloons compare to the density of air particles outside the balloons?

   _____

   _____

2. How does the air density inside a balloon change as it rises from the ground to the air?

   _____

   _____

1. The density of the air inside the balloons is lower than the density of the air outside the balloons.

2. As the air is heated, the air particles move more quickly and they spread out.

This continues until the heated air inside the balloon becomes less dense than the air outside, and the balloon rises.

How are the elements organized?

To understand the periodic table, study the key. It will tell you the meaning of each symbol and number.

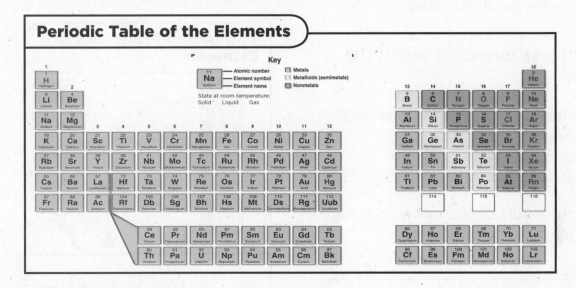

Periodic Table of the Elements

Answer these questions about the table.

1. Look at the key. What are the three main categories of elements?

_____

2. What is the symbol for sodium? What is its atomic number?

_____

3. In what other ways are the elements arranged?

_____

_____

1. The three main categories of elements are metals, metalloids, and nonmetals.

2. The symbol is Na, and its atomic number is 11.

3. The elements are arranged in numerical order by atomic number. They are also arranged by their states of matter: solids, liquids, and gases.

Name _____   Date _____

How do scientists use the periodic table?

This diagram shows ways that elements are grouped together. Notice that some are stacked into columns and others are linked into rows.

Answer these questions about the diagram.

1. Which group of elements is magnetic?

_____

2. Which group of elements is most reactive?

_____

3. How are elements classified into groups?

_____

_____

Comparing Properties of Elements

Bananas have lots of potassium (K). Elements in the potassium column react with nonmetals.

Many nails are made of iron (Fe), cobalt (Co), or nickel (Ni). These elements are all magnetic.

This is a sample of fluorine (F). All the elements in the flourine column form salts with elements in column 1.

1. The middle group, containing iron (Fe), cobalt (Co), and nickel (Ni), is magnetic.

2. Elements in the first group are most reactive.

3. Elements are classified into groups according to their properties, such as reactivity.

How does matter change state?

The diagram below shows the different states of water. The insets show the spacing and movement of water particles.

How Water Changes State

solid — Ice melts when energy is added. The particles move faster.

liquid — As energy is added to liquid water, the particles move faster. Some turn to gas.

gas — Water vapor is a gas. Its particles move very fast.

Answer these questions about the diagram.

1. What states of matter are shown?

_____

2. How are the particles in these three states different?

_____

_____

_____

_____

1. solid, liquid, and gas.

2. The particles in solid ice are close together and move slowly. The particles in liquid water are far apart and move quickly. The particles in gaseous water vapor are spaced farther apart and move fastest.

Name _____ Date _____

What are chemical changes?

The diagram below illustrates how iron reacts with sulfur. Look for differences between the two substances before and after the reaction.

Answer these questions about the diagram.

1. Which of the original substances is a metal?

   _____

   _____

2. What kind of change is being shown in this diagram? How do you know?

   _____

   _____

   _____

   _____

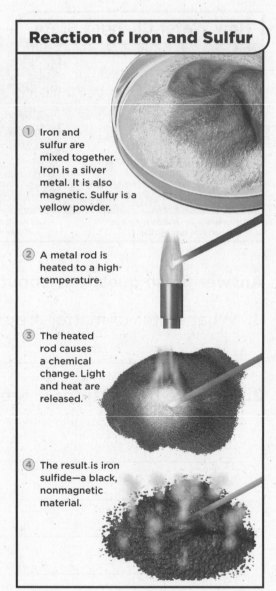

Reaction of Iron and Sulfur

① Iron and sulfur are mixed together. Iron is a silver metal. It is also magnetic. Sulfur is a yellow powder.

② A metal rod is heated to a high temperature.

③ The heated rod causes a chemical change. Light and heat are released.

④ The result is iron sulfide—a black, nonmagnetic material.

1. iron
2. A chemical change is shown. Heat energy was used to create one new substance—iron sulfide—from two substances—iron and sulfur.

What is a mixture?

Look at the photographs and labels to learn about four different types of mixtures.

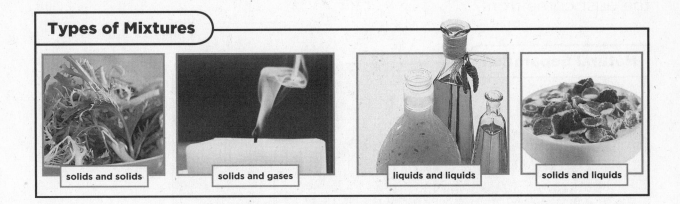

Types of Mixtures

| solids and solids | solids and gases | liquids and liquids | solids and liquids |

Answer these questions about the photographs.

1. What are three ways solids can form mixtures?

_____

2. What liquids are mixed in the third photo?

_____

3. What type of mixture is salt water?

_____

1. They can be combined with other solids, with gases, or with liquids.
2. oil and vinegar
3. It is a solid in a liquid.

How can you separate the parts of a mixture?

This photograph shows a car covered with dust. Where has the dust come from?

Natural Separation

Answer these questions about the photograph.

1. What substances make up this mixture?

   _____

2. What is causing the settling of this mixture?

   _____

   _____

1. air and dust
2. The dust is heavier than the air, so the dust eventually comes down out of the air and lands on a surface.

What are compounds?

Read the equations from left to right to learn what is produced when two elements are combined.

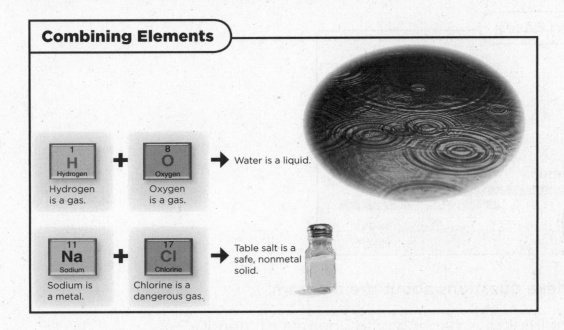

Combining Elements

| 1 H Hydrogen | **+** | 8 O Oxygen | **→** | Water is a liquid. |

Hydrogen is a gas.   Oxygen is a gas.

| 11 Na Sodium | **+** | 17 Cl Chlorine | **→** | Table salt is a safe, nonmetal solid. |

Sodium is a metal.   Chlorine is a dangerous gas.

Answer these questions about the diagram.

1. What compound is formed when hydrogen and oxygen are combined?

   _____

2. What must take place in order for two gases to combine to form a liquid?

   _____

3. How do you know that a chemical reaction occurs when sodium and chlorine are combined?

   _____

   _____

   _____

1. water
2. a chemical reaction
3. Chlorine is a gas, and sodium is a metal. Table salt is neither a gas nor a metal, so the elements have lost their original properties. This signifies a chemical change.

© Macmillan/McGraw-Hill

Name _____ Date _____

What are acids and bases?

Blue litmus paper turns red if it is touched by an acid. Red litmus paper turns blue when it is touched by a base.

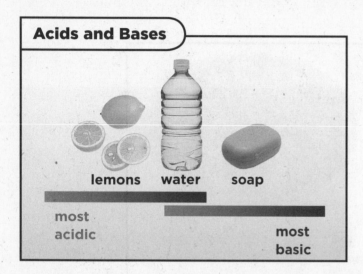

Acids and Bases

lemons water soap

most
acidic

most
basic

Answer these questions about the diagram.

1. What color litmus paper would you use to identify a lemon? What color does it turn the litmus paper?

_____

_____

2. What color litmus paper would you use to identify soap? What color does it turn the litmus paper? Why?

_____

_____

3. Would water turn red litmus paper blue? Why or why not?

_____

_____

© Macmillan/McGraw-Hill

How do forces change motion?

The speed skaters in this photograph are turning a corner.
Think about how you could describe their motion.

Acceleration

Answer these questions about the photograph.

1. Explain why turning is a form of acceleration.

_____

2. What force keeps these skaters moving? What must happen to stop their motion?

_____

1. Turning causes a change in direction.   2. Inertia keeps them moving. A force
                                                must be applied to stop their motion.

© Macmillan/McGraw-Hill

Name _____ Date _____

What is gravity?

This diagram illustrates apples falling from a tree. Gravity acts over a distance and pulls all objects together.

The Effect of Gravity

Answer these questions about the diagram.

1. What does the pull of gravity depend on?

   _____

   _____

2. Will Earth's gravity act on different-sized apples in the same way?

   _____

   _____

3. If the Sun's gravity is stronger, why aren't the apples pulled out toward outer space?

   _____

   _____

1. the amount of matter and the distance between objects
2. No, because objects with more mass pull together stronger than objects with less mass. Earth's gravity pulls stronger on larger apples.
3. The Sun is very far away, so the apples respond to Earth's gravity.

How do forces affect motion?

The numbers below each item represent the newtons of force needed to lift each item in the backpack.

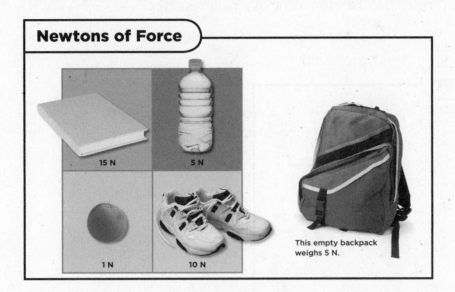

Newtons of Force

15 N

5 N

1 N

10 N

This empty backpack weighs 5 N.

Answer these questions about the diagram.

1. Which item would you need the most force to lift? Which item would you need the least force to lift?

   _____

2. If you put all of the items in the backpack, how much force would you need in order to lift the backpack?

   _____

   _____

3. If the shoes were removed, how much force would you need in order to lift the backpack?

   _____

1. the book; the orange
2. 36 N
3. 26 N

Name _____ Date _____

How do forces affect acceleration?

To read this diagram, study the sizes of the pumpkins and the number of people pulling them. The top arrow represents acceleration, and the bottom arrow represents the amount of force.

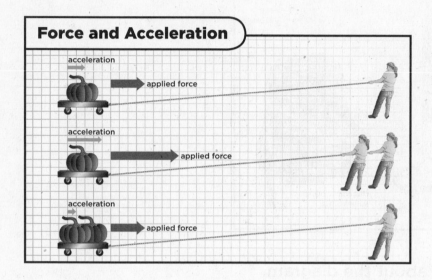

Force and Acceleration

acceleration
applied force

acceleration
applied force

acceleration
applied force

Answer these questions about the diagram.

1. How are the mass, force, and distance traveled different in the first and last illustrations?

_____

_____

2. How are mass, force, and distance traveled illustrated in the second drawing?

_____

_____

1. The force is the same in both illustrations. There is more mass in the third illustration, so less distance is traveled in the third illustration.

2. In the second illustration, two people are pulling the same amount of mass. The same mass with twice the force means twice the distance traveled.

What is work?

This diagram shows three different points on a roller coaster. Study the diagram to understand the different types of work being illustrated at each point.

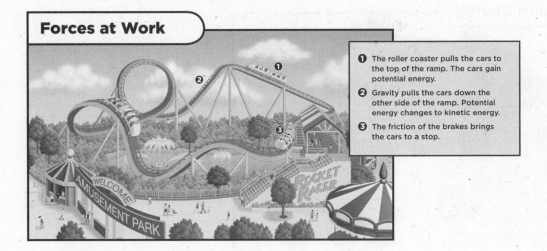

Forces at Work

❶ The roller coaster pulls the cars to the top of the ramp. The cars gain potential energy.

❷ Gravity pulls the cars down the other side of the ramp. Potential energy changes to kinetic energy.

❸ The friction of the brakes brings the cars to a stop.

Answer these questions about the diagram.

1. What is the source of energy that pulls the roller coaster to the top of the ramp?

_____

2. What is the source of energy that moves the roller coaster down the ramp? What is the source of energy that moves the roller coaster from the ramp to the end of the ride?

_____

3. Would there be any friction if the cars did not have brakes?

_____

1. a motor in the roller coaster

2. gravity; kinetic energy

3. Yes, because even smooth surfaces have a little friction.

How can energy change?

This house uses solar panels to collect light energy. Read the captions to understand how energy is transformed in this house.

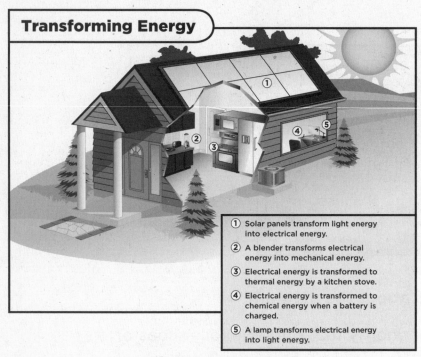

Transforming Energy

① Solar panels transform light energy into electrical energy.

② A blender transforms electrical energy into mechanical energy.

③ Electrical energy is transformed to thermal energy by a kitchen stove.

④ Electrical energy is transformed to chemical energy when a battery is charged.

⑤ A lamp transforms electrical energy into light energy.

Answer these questions about the diagram.

1. In what form does the energy enter the home?

_____

2. What type of energy transformation occurs first?

_____

3. What are the other forms of energy into which electrical energy is transformed?

_____

1. It enters as light energy.
2. The light energy is transformed into electrical energy.
3. Electrical energy is transformed into light energy, thermal energy, chemical energy, and mechanical energy.

What are two other simple machines?

The diagram below shows information about some pulleys. It displays the direction in which force is applied to the pulley, and the direction in which the load attached the pulley moves.

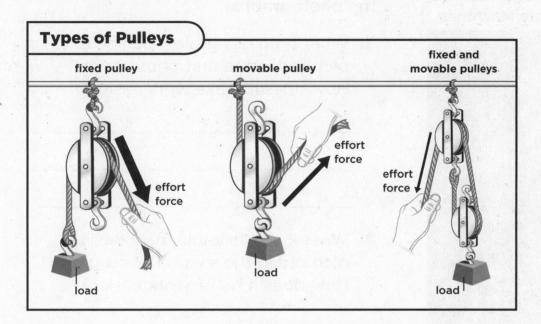

Types of Pulleys

fixed pulley   movable pulley   fixed and movable pulleys

effort force   effort force   effort force

load   load   load

Answer these questions about the diagram.

1. What type of pulley is shown in each drawing?

_____

_____

2. How is a movable pulley different from a fixed pulley?

_____

_____

1. The first drawing shows a fixed pulley, the second drawing shows a movable pulley, and the third drawing shows fixed and movable pulleys.

2. In a movable pulley, the force is being applied in the same direction in which the load is moving.

What are inclined planes?

Each of these photographs shows a different type of simple machine. Study the photographs to understand how they work.

Three Simple Machines

ramp

screw

wedge

Answer these questions about the photographs.

1. What type of simple machine is pictured in the first photograph? How does it make work easier?

_____

_____

_____

2. What kind of simple machine is pictured in the second photograph? How does it make work easier?

_____

_____

_____

3. What kind of simple machine is pictured in the third photograph? How does it make work easier?

_____

_____

1. An inclined plane is pictured; it supports most of an object's weight, making it easier to move.

2. An inclined plane is pictured; its twisted inclined plane structure allows you to apply a small force to get it to move

easily through tough material.

3. A lever is pictured; it allows you to apply a small force to one end of the lever, making it apply a stronger force at the other end, using that to slice the apple.

What is heat?

On this thermometer, the *C* on the top left stands for degrees *Celsius*, while the *F* on the far right stands for degrees *Fahrenheit*.

Measuring Temperature

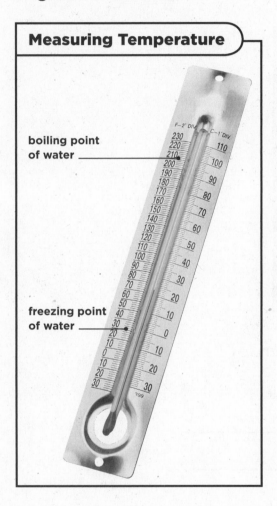

boiling point of water

freezing point of water

Answer these questions about the photograph.

1. What does this instrument measure? How does it work?

_____

_____

_____

2. What is the freezing point of water in degrees Fahrenheit and in degrees Celsius?

_____

_____

_____

1. A thermometer measures temperature. Most thermometers contain a liquid, such as alcohol. When it is warmed, it expands and rises in a glass tube.

The markings on the side indicate temperature.

2. The freezing point of water is 32°F and 0°C.

Name _____ Date _____

How does heat travel?

This diagram shows how heat is transferred from the flame to the water.

Heat Transfer

Answer these questions about the diagram.

1. Where is conduction occurring? Explain.

   _____

   _____

2. Where is convection occurring? Explain.

   _____

   _____

3. How do the particles in the water change as they are heated?

   _____

   _____

1. Conduction takes place at the bottom of the pot, where the thermal energy of the flame is transferred to the pot.

2. Convection takes place in the water, because heat energy is transferred throughout the water.

3. As the water is heated, the particles move faster, become less dense, and move farther apart.

© Macmillan/McGraw-Hill

How does sound travel?

The speedometer in the diagram below is used to show how fast sound travels through different substances.

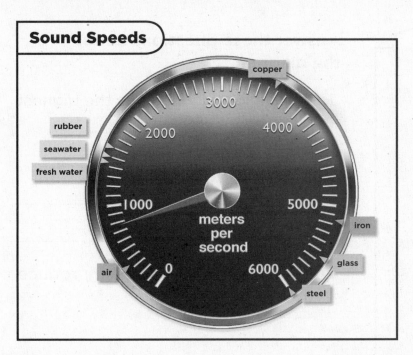

Sound Speeds

copper
3000
rubber   4000
2000
seawater
fresh water
1000   5000
meters
per   iron
second
glass
air   0   6000
steel

Answer these questions about the diagram.

1. Does sound travel faster through glass or steel? How much faster?

   _____

   _____

2. Through which of the following does sound travel fastest: solids, liquids, or gases? Why?

   _____

   _____

1. Sound travels faster through steel (at a rate of 450 meters per second).
2. Sound travels fastest through solids. Because sound waves are areas of compressed particles, denser materials transmit sound faster.

Name _____   Date _____

How do sounds differ?

To understand the differences in sounds, look at the
wavelength, frequency, and volume of sound waves.

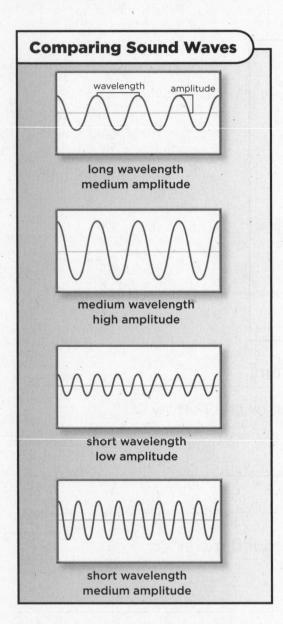

Comparing Sound Waves

wavelength   amplitude

**long wavelength
medium amplitude**

**medium wavelength
high amplitude**

**short wavelength
low amplitude**

**short wavelength
medium amplitude**

**Answer these questions about
the diagram.**

1. Which sound wave has the highest
pitch?

_____

_____

_____

2. Which sound wave would produce
the quietest sound?

_____

_____

3. Which sound wave would produce
the loudest sound?

_____

_____

1. Waves that have a very short
wavelength have a high pitch, the short
wavelength medium amplitude.

2. The long-wavelength, medium-
amplitude wave would most likely
produce the quietest sound.

3. The loudest sound would be produced
by the short-wavelength, medium-
amplitude wave.

Use with **Lesson 2
Sound**

What is light?

This diagram shows the wavelengths of the three colors in a stop light—red, yellow, and green.

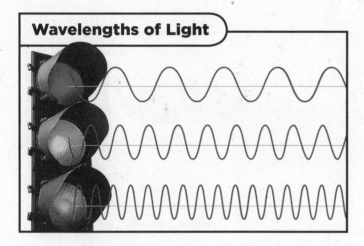

Wavelengths of Light

Answer these questions about the diagram.

1. Which color has the shortest wavelength?

_____

2. How would you expect the wavelength for orange to appear? Why?

_____

_____

3. Are all light waves visible? Explain.

_____

_____

© Macmillan/McGraw-Hill

Name _____ Date _____

How does light travel?

This diagram shows the path of light as it enters the eye and continues to the optic nerve.

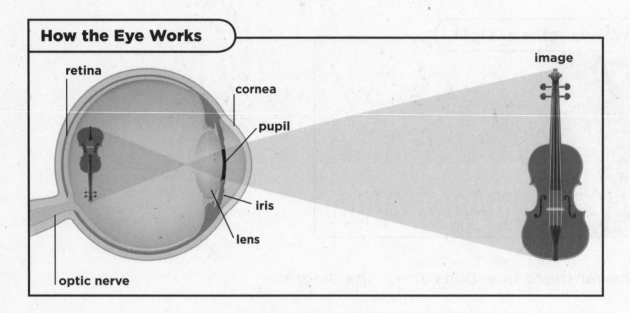

How the Eye Works

retina

cornea

pupil

image

iris

lens

optic nerve

Answer these questions about the diagram.

1. What is the opening in the iris that allows light in?

_____

2. How do images appear on the retina?

_____

_____

3. How do images get transmitted to the brain?

_____

_____

1. the pupil
2. Images appear upside down on the retina.

3. Nerves in your eyes, called optic nerves, send the images to your brain.

How do charges move?

This diagram shows the flow of electricity in the flashlight's circuit when the switch is turned on and when it is turned off.

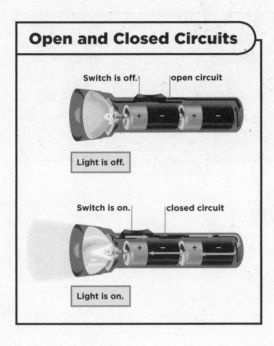

Open and Closed Circuits

Switch is off.   open circuit

Light is off.

Switch is on.   closed circuit

Light is on.

Answer these questions about the diagram.

1. Look at the top diagram. What happens when the switch is off?

_____

_____

_____

_____

2. How is the bottom diagram different from the top diagram?

_____

_____

_____

_____

1. When the switch is off, the circuit is open. Since the circuit is open, no current will flow through it, and the light will not turn on.

2. In the bottom diagram, the switch is on, so the circuit is closed. The current flows through the circuit and lights the light bulb.

Name _____ Date _____

What are series and parallel circuits?

This diagram shows two types of circuits in opened and closed positions.

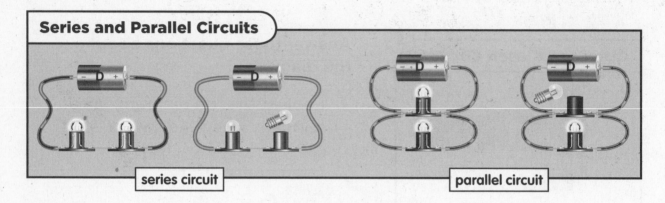

Series and Parallel Circuits

series circuit

parallel circuit

Answer these questions about the diagram.

1. What happens in the series circuit when a light bulb is removed? Why?

_____

_____

2. What happens in the parallel circuit when a light bulb is removed?

_____

_____

1. The remaining light bulb will not be lit because an open circuit is created.

2. Each object is connected to the power source through a separate path. Even with one bulb removed, it is still part of a complete circuit.

What is an electromagnet?

This diagram shows how an electromagnet powers a motor.
It shows how the parts of the motor work together.

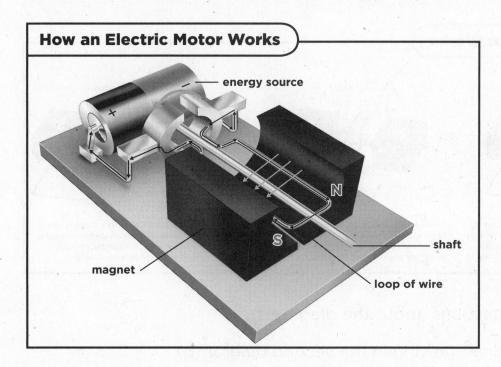

How an Electric Motor Works

- energy source
- N
- S
- magnet
- shaft
- loop of wire

Answer these questions about the diagram.

1. What is the coil of wire attached to?

_____

2. What causes the shaft to turn? How is energy changed
as a result?

_____

_____

_____

1. The coil is attached to the motor shaft.
2. The interaction of the magnetic field of
the wire and permanent magnet causes
the shaft to turn. Electrical energy is
changed into mechanical energy.

Name _____ Date _____

What is a generator?

These diagrams show a generator. Read the caption below each diagram to understand how a generator works.

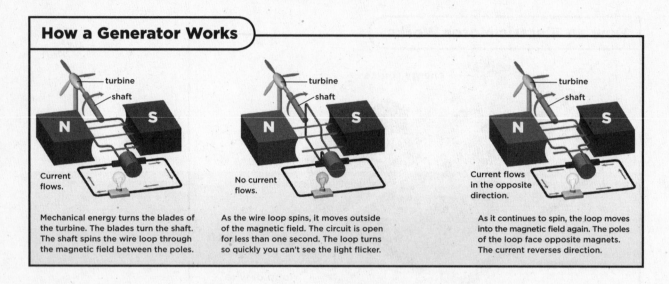

How a Generator Works

turbine
shaft
N   S
Current flows.

Mechanical energy turns the blades of the turbine. The blades turn the shaft. The shaft spins the wire loop through the magnetic field between the poles.

turbine
shaft
N   S
No current flows.

As the wire loop spins, it moves outside of the magnetic field. The circuit is open for less than one second. The loop turns so quickly you can't see the light flicker.

turbine
shaft
N   S
Current flows in the opposite direction.

As it continues to spin, the loop moves into the magnetic field again. The poles of the loop face opposite magnets. The current reverses direction.

Answer these questions about the diagrams.

1. Why is no current flowing in the second diagram?

_____

_____

2. Why does the current flow in the opposite direction in the third diagram?

_____

_____

1. When neither side of the loop is in the magnetic field, no current flows.

2. The loop passes down through the field as it continues to turn, causing the current to flow in the opposite direction.